A Sprint to the Finish

Russell McGuire

Published by SDG Strategy, LLC, 2020.

A SPRINT TO THE FINISH

First edition. October 1, 2020.

ISBN: 978-1393924012

Written by Russell McGuire.

Table of Contents

Preface

———

E arlier this year T-Mobile finally completed the acquisition of Sprint. In the days that have followed I've seen many social media posts — some celebrating this hard fought conclusion, others reflecting on the bittersweet end to one of the most innovative service providers in U.S. telecom history. I've received many notifications from LinkedIn to congratulate friends and contacts for their "new position" in the combined company. Congratulations to you all!

But in this book, I want to reflect back a bit on some strategic moments in the company's history. For those that don't know, I worked for Sprint as a strategy executive from 2003–2014, but my ties to the company go way back to the 1920s. My grandfather Carl Spaid started working in telecom in 1913 in his final year of high school as a lineman's assistant at the local telephone company in Joplin, Missouri. He continued with the company while an engineering student at the University of Missouri. In 1917 he enlisted in the signal corps of the U.S. Army. When the war ended in 1918 he was released from his service commitment and returned to the phone company. In 1925 he moved to Abilene, Kansas and joined United Telephone, the company that would become Sprint. In March 1929 (the month my mother was born) he moved his family to Kansas City and became Chief Engineer. He would later rise to become President of United Telephone of Kansas and Missouri (about one third of the entire United Telephone operation). So, it's natural that I've had a healthy interest in the history of the company, and especially the strategic decisions made over the 121 years of its existence.

In the coming pages I want to share some stories from that history. My grandfather passed away before I was born, so some of the stories I only know from a distance. Similarly, although I was involved when we took a swing at merging with T-Mobile in 2014, I had no involvement or inside perspective on the activities that led up the final deal, so that story I'll tell from a pretty detached perspective. But there were a number of key strategic transitions that happened while I was at Sprint. I'll tell these stories very much from my perspective. That means the lens that I was looking through and my own opinions will undoubtedly color them. Other people likely will remember the same periods differently. I don't claim that I had complete visibility into everything that was happening (far from it), therefore my perspectives may not be completely accurate. But I hope that the stories will be entertaining and informative, especially for those wanting to learn about strategic decision-making.

Over the coming months and years I plan on sharing more and more about tools and approaches that can help with strategic decision-making at all kinds of organizations, so if you find this interesting, please follow my blog at clearpurpose.media[1].

---

1. http://clearpurpose.media/

A Sprint from the Start

———

A rguably the first strategic decision in Sprint's history was the decision by the company's founder to enter the telecom business.

Alexander Graham Bell invented and patented the telephone in 1876 and the Bell Telephone Company began wiring every major city for telephone service. When Bell's patent expired, entrepreneurs started building out networks in unserved cities and sometimes establishing rival networks (at a lower price) in cities already served by Bell.

Cleyson L. Brown was a serial entrepreneur in Abilene, Kansas. He and his brother Jacob had already started a water company. They used water power to start an electric company. And then they used their electric poles to start a telephone company. They began offering telephone service in 1899 and the Brown Telephone Company was officially chartered in 1902.

The second major strategic decision the company made was to grow through consolidation. By 1911 the company served over 4,000 telephones in the Abilene area. That year Brown led the consolidation of other independent (not Bell) telephone companies in Kansas. The newly combined companies renamed themselves United Telephone Company. Over time the company acquired 68 water, electric, and telephone companies and in 1925 formed a holding company called United Telephone and Electric (UT&E).

The depression was hard on many companies. United managed to survive until 1934 when a number of factors combined to cause the company to file bankruptcy, but by 1937 United had recovered and was able to emerge from bankruptcy and continue to grow. In 1939 the holding company changed names to United Utilities, Incorporated.

Brown had hired Skip Scupin in 1921 and Scupin led many of the company's significant projects over the years. In 1959, Scupin became president and moved the company's headquarters to Kansas City.

In the preface I mentioned that my grandfather worked for United. To provide a sense for the size and structure of the company during this period I have copies of portions of the annual reports from 1947 and 1954.

As of December 31, 1947, United served 155,103 telephone lines, had 8,492 electric customers, 2,882 gas customers, 414 water customers, and had divisions focused on Petroleum and Merchandising. The telephone operations were broken into three divisions: the Eastern Group, the Central Group, and the Western Group, each of which consisted of separate operating companies. For example, my grandfather was CEO of the Western Group comprised of American Telephone Company (26,637 lines), Nodaway Telephone Company (2,390 lines), Siloam Springs Telephone Company (914 lines), and United Telephone Company of Missouri (23,361 lines).

By December 31, 1954 the company had more than doubled to serve 358,213 telephone lines. The non-telecom utilities had been consolidated under the Central Kansas Power Company with Scupin as president with 11,291 electric customers, 4,400 gas customers, and 568 water customers. The telecom business was now in four divisions with my grandfather serving as president of the division containing United Telephone of Missouri (53k lines), United Telephone of Kansas (36k), The McKrae Telephone Company (5k), Arkansas Associated Telephone (3k) and Siloam Springs Telephone (2k).

One of the men Skip Scupin was able to recruit to the new headquarters in Kansas City was Paul Henson. Henson had started his telecom career climbing poles for Lincoln Telephone & Telegraph while attending the University of Nebraska. In 1964 Scupin retired and Henson succeeded him as President of United Utilities. Henson would become perhaps United/Sprint's greatest leader.

Henson continued the long history of the company's growth through consolidation, launching an ambitious campaign named "Growth Through Acquisitions" with the goal of reaching 2 million telephone lines and $1 billion in assets. By 1966 Henson had done deals in Ohio, Oregon, Tennessee, Virginia, and Pennsylvania to double the size of the company. Over the next three years he did more deals in Florida, Texas, North Carolina, and South Carolina to double the size of the company again!

While growing the telecom operations, Henson also was divesting the non-telecom assets and in 1972 the company name was changed again, this time to United Telecommunications, Inc. (or United Telecom). The company began experimenting with various related businesses, some of which we will talk about in the coming days. During these strategic experiments Henson recruited Bill Esrey to join the company as Vice President of Finance. In 1984 Esrey succeeded Henson as President of United Telecom.

Esrey continued to grow the traditional telephone company through acquisitions. The largest and most strategic was the acquisition of Centel (Central Telephone Company) in 1993. At the time United was the third largest telephone company in the U.S. (behind AT&T and GTE) and Centel was the fifth largest. (The fourth largest, Continental Telephone would later be acquired by GTE.) Centel brought 1.5 million telephone lines to the company, but also brought strategic growth markets being the monopoly provider in Las Vegas, several Chicago suburbs and other growth markets in the southeast.

The traditional local telephone business started by the Brown brothers in 1899 continued to be a core asset of the company until the merger with Nextel in 2005. Following that merger the local assets were spun off as Embarq which continued to participate in the consolidation of the independent local telecom industry and today is part of CenturyLink serving over 5 million local customers with $23 billion in revenue.

A Sprint for the (Long) Distance

M y role in corporate strategy at Sprint included managing the incredible resource which was the Corporate Research Center. During (relatively) quiet periods I would head down to the "stacks" and would take the time to dig through old company records and annual reports to understand strategic decisions that had been made in the past. Tracking the thinking behind the entry into the Long Distance market was particularly fascinating.

As we discussed in the last chapter, by the 1970s United Utilities/ Telecom had become the 3rd largest telephone company in the country (behind AT&T and GTE) serving millions of customers across the country. However, United didn't really compete with AT&T, GTE, Continental, Centel or any of the other independent telephone companies in the market. Telecom was a natural (and heavily regulated) monopoly with each geography being served by a single provider. That provider had made significant capital investments over the decades to establish the infrastructure needed to provide local service and to connect customers to AT&T Long Lines for long distance calls.

That started to change in the 1970s as the Department of Justice began to pursue an anti-trust lawsuit against AT&T. That lawsuit was filed in 1974 and reached its conclusion in 1982 with AT&T's agreement to break itself apart and open the long distance market to competition. You might think that companies like GTE and United would be thrilled that the government was going after the industry leader, but you'd be wrong. A regulated monopoly can be a very attractive business. If the government messed with that arrangement for AT&T, it probably wouldn't be good for United or GTE either.

To understand this drive for change, we actually have to go back to the early 1960s. In 1963 a group of two-way radio salesmen got the idea to build a microwave radio network to connect truckers to their home offices. They called their company Microwave Communications, Inc. or MCI. Although constantly short of cash and weak on execution, those modest ambitions began to grow as the company sought the licenses needed from the Federal Communications Commission (FCC) to operate a microwave network. AT&T brought its lobbying weight to bear to block this potential competitor. In order to gain public and FCC support for their plan, MCI began to position their proposal against AT&T's monopoly TELPAK private line service, saying that they could provide better and cheaper service for transporting voice and data than AT&T, and provide it to smaller businesses being underserved by Ma Bell.

In 1967 MCI won their first major victory with the FCC, gaining a license to build a microwave network from Chicago to St. Louis and being deemed a common carrier with full rights to connect with other telecom networks. MCI then began the fight to build a nationwide microwave network and become a full-on competitor with AT&T for dedicated private line services to business customers. It is quite possible that, if AT&T had ignored MCI's original petition to the FCC, the tiny, underfunded, poorly managed company would have quietly vanished from the scene. But AT&T's heavy-handed attempts to crush this upstart raised red flags in Washington and across the country; red flags that eventually led the Justice Department to begin their anti-trust investigation of Ma Bell.

Switching back to United/Sprint, it is fascinating to me to hear the story play out in the very words of the company's leadership. Below I have provided quotes from the letters from United's Chairman Paul Henson "to our stockholders and employees" in each issue of the company's Annual Report. Most of these letters had two names attached — Henson and whoever was president at the time, but through this whole journey Paul Henson's stamp is clearly on the story. For the extended quotes below I will merely preface each with the Annual Report year. Keep in mind that the Annual Report comes out near the beginning of the following year, so, for example, the first quote from the 1973 Annual Report would have been issued early in 1974.

As I worked my way through these Annual Reports it was fascinating to see the strategic development process unfold. United had a well-established business strategy. They were constantly monitoring the external environment (regulatory, competitive, technology, economy, customer needs, etc.) As the environment started to shift, the company first sought to influence the external environment to maintain a status quo that served them well. When it became apparent that change was inevitable, they began to seriously evaluate external opportunities and threats and internal strengths and weaknesses to determine a new strategy. They pursued a portfolio approach managing investments in various businesses. Some failed quickly and were eliminated. One rose to the top as the future of the company. Company resources were managed to maximize the benefits to all stakeholders. Existing businesses helped fund investment in new ventures and provided the operational and customer base foundation for the success of these startups. See if you hear the same things as me in the "voice" of Paul Henson as you read his annual updates below.

1973: "Hopefully, in 1974 we will see some resolution to the complex and difficult problems that beset the telephone industry with the advent of structured competition. What started out as a well-intentioned attempt by the Federal Communications Commission to introduce limited competition into the telephone industry has turned into a nightmare of polarized industry positions, conflicting regulatory decisions and confused legal actions. The only profit thus far has gone to a few major communications users and to the legal profession, and that largely at the expense of the general rate-paying public."

1974: "The concept and definition of competition in the telephone industry, as postulated by the Federal Communications Commission, has been difficult to understand — at best. ... The timing, the allegations and the relief sought by the Justice Department's antitrust suit against AT&T merely add to the confusion. Disposition of the case against AT&T will be painfully slow, and its impact on United Telecom remains uncertain... Fundamentally, questions relating to competition and interconnection are matters of national policy which must be resolved by the Congress. We are hopeful that the AT&T antitrust suit will serve to hasten the day when Congress will either reaffirm or deny that previous national policies on communications have been and are in the public interest. These policies have given the United States the best telephone service in the world at the lowest relative cost to the public it serves."

1975: "It became more apparent in 1975 that the threatened erosion of the telephone industry's integrated approach to providing universal telephone service was becoming a reality. Contrived competition, sponsored by the Federal Communications Commission and introduced without consideration of the economic impact on residential telephone service, established roots in the intercity communications market. ... Such unnecessary duplication of established telephone company intercity private line services will be detrimental to the long-term interests of our residential and small business customers who will bear the burden of accelerated increases in price of basic telephone service. Ultimately, the effect could be harmful to the employees and stockholders of United Telecom. ... In 1976, United Telecom, with more than 1,500 other telephone companies, will make a determined effort to enlist the support of its employees, the customers it serves, its stockholders and its congressional representatives to amend the Communications Act of 1934. The purpose will be to reaffirm the explicit intent of the original legislation's national goal of providing '...a rapid, efficient, nationwide and worldwide wire and radio communications service with adequate facilities at reasonable charges.'"

1976: "A significant development in 1976 was the recognition by the U.S. Congress that its attention should be focused on reexamining national telecommunications policy. ... Exploratory hearings were held in the U.S. House of Representatives in 1976 on the impact of competition in the domestic telecommunications industry. In its testimony, United Telecom stressed the need for Congress to provide definition and direction in the form of a national telecommunications policy."

1977: "No recap of 1977 would be complete without mention of the industry's effort to have the United States Congress examine and revise the national telecommunications policy."

1978: "New emphasis was placed on corporate long-range planning in 1978, with intensive study of our existing markets as well as other service-oriented markets related to United's present business activities. The planning process is complicated by lack of a clear-cut national telecommunications policy. It is evident, however, that the policy evolving in Congress will call for more competition in both the inter-city and terminal equipment segments of the telephone business. ... A revised draft of a bill rewriting the Communications Act of 1934 is to be introduced in the House of Representatives early in 1979. Legislation also may be initiated in the Senate. This provides hope that the formulation of a national telecommunications policy can be completed by 1980."

1979: "The year 1979 marked both the end of a decade of solid growth for United Telecom and, more important, the beginning of a new chapter in the company's history. ... It was a year of intense planning for United Telecom's direction in the 1980s. We systematically analyzed both the strategic opportunities presented by changes in the business environment in which the telephone industry operates and the changes needed by United companies to take full advantage of new technologies and markets. ... Congressional action on national telecommunications policy could occur in 1980. ... United Telecom enters the 1980s with a successful record as a regulated telephone company and established bases in competitive telecommunications and computer services markets. We firmly believe a significant portion of the corporation's future growth will come from non-regulated, competitive businesses. Our goal for the 1980s is to become as strong in competitive businesses as we are in the regulated telephone industry. ... This calls for a major redirection of the corporation's activities and resources in which we will consolidate our regulated telephone operations, build competitive telecommunications businesses into a major second sector and develop the computer businesses into a third major entity."

1980: "On a national level, a few faltering steps were taken in 1980 toward what eventually will be a less regulated, more competitive environment for the telecommunications industry. ... It is difficult for United Telecom, or any other participant in this rapidly growing industry, to conduct meaningful long-range planning and development activities when the ground rules remain subject to change."

1981: "United Telecom had a very good year in 1981. Our company successfully broadened its scope of operations and strengthened its competitive position while recording a solid earnings performance. ... Our strategic direction for the decade of the '80s became more evident in a series of 1981 developments. ... Our major development efforts in 1982 will be working toward our long-term objective of providing enhanced voice and digital services via an intercity network. We will build this network on the capabilities and facilities of UNINET and ISACOMM. ... Completion of this network will give us the capability of offering selected enhanced services for the office and the home, both inside and outside the areas now served by our telephone companies. The enhanced services market is large and growing. We are confident our corporate resources and capabilities equip us to serve this market successfully. ... Early in 1982, we created a fourth operating group to concentrate our management and capital resources in further developing our network capabilities. William T. Esrey, who had been executive vice president and chief financial officer of United Telecom, was named president of the new operating group. ... The momentous announcement in early 1982 of an agreement to settle the government's antitrust suit against American Telephone and Telegraph Company will have significant impact on national telecommunications policy."

1982: "United Telecom had a difficult year in 1982 — preparing for the fundamental changes unfolding in the telecommunications industry while coping with a recession which slowed or reversed historical growth patterns. ... Our major new thrust in 1982 was the formation of a fourth operating group, United Telecom Communications, Inc. This action followed the January acquisition of Insurance Systems of America and its controlling interest in ISACOMM, a satellite communications company. United Telecom Communications links ISACOMM with the packet switching services of UNINET to offer a broad range of digital voice and data services."

1983: "Our foundation business is, and will continue to be, providing local network services. We intend to protect and enhance our investment in the local exchange telephone network, strengthening our position as a low-cost provider and adding the innovative services our customers want. ... We also have initiated a major venture thrust in integrated intercity services. Our current emphasis is on value-added network services and video teleconferencing. We intend to make the necessary major investments over the next decade to build these intercity services into another profitable core business. ... Our December 31, 1983, agreement to acquire U.S. Telephone, a rapidly growing participant in the long distance telephone market, further demonstrates our commitment to developing a full service intercity communications capability. We're confident we have the market knowledge, the resources and the skills required to succeed."

1984: "As anticipated, 1984 was a memorable year for the telecommunications industry and an eventful year for United Telecom. ... We have continued to prepare United Telecom for the future. Our core telephone business remains very strong and is well positioned for the future. Still, we recognize that the opportunities for future earnings gains from telephone operations are limited. ... With the deregulation of the mature telephone industry, our future investment in these properties will grow more slowly. While we will protect and maintain our position as the low-cost provider of local network services, introduce new services and meet competition head-on, we cannot expect to match our historical earnings growth rates in telephone operations. ... Our primary development thrust remains in integrated intercity communications services. Our greatest challenges and our greatest opportunities lie in developing this business as United Telecom's second core business. ... Great strides were made in 1984. We acquired U.S. Telephone, a major long distance company, and integrated its operations with those of ISACOMM and UNINET. We now have a full spectrum of services to offer our customers. ... We also established a common identity for these companies as we pulled our intercity operations together in a single organization. Beginning in January 1985, we offered our services under the name of US Telecom. ... Beyond that, we initiated action in 1984 to build a 23,000-mile nationwide digital communications network that will give US Telecom a service capability second to none. In many respects we exceeded our own expectations. ... Because of the scope and timing of our network construction plans, which call for spending in excess of $700 million in both 1985 and 1986, we are seeking a partner or partners to join us in developing our network. ... We've had a full year of experience in the new telecommunications operating environment which came with the January 1984 break-up of the Bell System. The course to a fully deregulated environment remains uncertain, yet manageable."

1985: "The year 1985 confirmed United Telecom's commitment to grow and to become a leader in the markets it serves. We see a bright future in telecommunications services. In 1985, we invested $1.1 billion to strengthen and increase our stake in that future. More than half of that investment went into our core telephone business, where we have earned an industry leadership position. From that solid base, we're rapidly building a second core business in intercity communications. As highlighted in this report, we are building the nation's most advanced telecommunications network — on schedule and below budget. ... Our network strategy was a key element in our ability to form a strong partnership. We announced our intention to form a partnership with GTE in January 1986 after several months of negotiations. It will combine our US Telecom long distance and data communications businesses with those of GTE Sprint and GTE Telenet to form US Sprint Communications Company. Our network strategy and our strong position in the corporate marketplace match well with Sprint's size and strength in the residential and small business marketplace."

1987: "For United Telecom, 1987 was a year of achievement, challenge, and disappointment. Although there was much to be encouraged about, the financial performance of US Sprint fell short of expectations. ... Without minimizing the disappointment or underestimating the challenge ahead, we are convinced that 1987 will be remembered as the year we positioned United Telecom for a truly rewarding future. The price we paid in 1987 to achieve that positioning with US Sprint was higher than we expected. However, it was not higher than we believe is warranted by the potential for substantial, long-term earnings growth. ... Every day we are moving closer to realizing our vision of US Sprint. The nation's only all-digital, all-fiber-optic network is the springboard for US Sprint's rapid ascension in the long-distance business."

1988: "The year 1988 should go on record as a watershed year in United Telecom's 89-year history. It was a year of extraordinary change and great progress. ... Achieving one milestone after another, United Telecom has moved to the forefront of the dynamic telecommunications industry. The most significant events included: Our purchase of a controlling interest in US Sprint from GTE. As of January, 1989, United Telecom owns 80.1 percent of US Sprint. ... Beyond such milestones, 1988 was a year when we started to see the positive results of our long-term strategy. ... US Sprint's abrupt turnaround in the last half of 1988 was widely recognized in the financial community and reflected in a healthy stock price gain. United Telecom's common stock had a year end close of $46 3/8 compared to a 1987 close of $24 5/8... an increase of nearly 90 percent. ... US Sprint's long and sometimes tortuous road to profitability never altered our vision of the future. ... When we decided to enter the long distance business near the beginning of this decade, we recognized it was a bold stroke. We knew entering this highly competitive market was a huge undertaking for a company our size. But we are convinced now, more than ever, that we were the right company, with the right strategy, at the right time. ... US Sprint has changed the nature and the direction of this corporation. We have made, and kept, our major promises at US Sprint. We have financed, built and moved millions of customers to the world's most technologically advanced network. ... Clearly we are re-emerging as a growth company. We're not focused on our size, and we hope we never are. The real payoff is what the increased ownership of US Sprint allows us to do for our customers, our shareholders and our employees. ... United Telecom never has been a 'me, too' organization. Our aim is to rise above the competition. Our goal is to become the best telecommunications company in the world. We made considerable progress toward that goal in 1988. And we expect to continue on that path in the coming year."

1989: "Our mission is straightforward — to be the best telecommunications company in the world — to rise above the competition and become the standard against which all others in the industry are measured. Our strong financial results for 1989 are one key measure of our progress. ... Reflecting the importance of US Sprint to the success of our company, we will ask shareholders to approve changing our name from United Telecom to Sprint Corporation when we exercise our option to purchase the remaining 19.9 percent interest in US Sprint from GTE. We believe all our companies will benefit from association with the widely recognized and respected Sprint name."

That, to me, wraps up an exciting and even emotional journey from a well managed but not overly aggressive regulated monopoly utility company to the dynamic, innovative, and fiercely competitive machine that many of us came to know and love as Sprint.

But before I close this chapter, I think it very meaningful to quote from one other letter in that 1989 Annual Report. This one was from Bill Esrey who had taken over the CEO role and was about to take over the Chairman role from Paul Henson:

"As we enter a new decade with confidence in our powerful local and long-distance divisions and our bold strategic plans, it's appropriate to pause and thank the man who had the foresight and courage to help us bring vision into reality... Paul Henson, who is retiring as our chairman of the board on April 17. When Paul arrived in 1959 at what was then United Utilities, the corporate staff consisted of the late Carl A. Scupin, president; Henson, his vice president; a clerk and two secretaries. United's annual revenues were $38.2 million that year. Although the company then served about 450,000 telephones, only 44 percent had local dial service [the other 56 percent needed to be connected by a switchboard operator] and only 10 percent had direct toll dialing [the ability to dial long distance without an operator's help]. Today, 31 years after Paul's arrival, United serves nearly 4 million local access lines, 81 percent of which have digital switching. With US Sprint, combined annual revenues have surged to more than $7.5 billion and assets are approaching $10 billion. ... Although Paul Henson embodies that most rare combination of business savvy and human warmth, he is much more than that. In a true sense of the word, he's a visionary... and a gentleman. He never has been one to take singular credit for collective achievement, but one wonders where this corporation would be without three decades of leadership marked by his unending quest for quality and innovation. His courage is only exceeded by his patience. In a world driven by short-term financial concerns, he held to long-term convictions, integrity and perseverance. ... Skeptics found it hard to believe that a relatively small Midwestern company could successfully take on the mammoth long-distance market... and do it on the basis of quality. It took a willful, yet humble man to set the course for this company to become a global force for positive change in telecommunications."

It remains an amazing story!

Sprinting into Wireless

In the last chapter we looked at how United Telecom expanded into Long Distance and became Sprint. The story involved joint ventures, acquisitions, and big bets. It was a long road with a few twists and turns. In this chapter we look at how Sprint became a wireless carrier, and it's more of the same — in fact much more.

The modern U.S. wireless industry dates back to the early 1980s when the FCC started distributing spectrum licenses for the new cellular wireless architecture. Those first licenses were given away — two in each market. One went to the local telephone company and the other went to an applicant who could prove they could build it out.

As a local telephone company, United received licenses in the areas where it served. Many of these areas were suburbs of larger cities. The FCC encouraged companies to partner together so that one wireless network could be built to serve an entire metro area. United Telephone formed a joint venture with other independent telcos and AT&T's Advanced Mobile Phone Service, Inc. (AMPS) subsidiary. The first round of licenses covered the largest cities in the country and United received a minority stake in the wireless operations in New York and Kansas City. In the second round, United gained positions in Orlando, four Ohio cities, Allentown, and Norfolk-Virginia Beach. In later rounds, the company continued to expand its footprint. When AT&T was broken up the AMPS operations were split between the 7 Bell Operating Companies which have now largely been reconsolidated into Verizon and (the new) AT&T.

By 1987 the wireless business, now named United TeleSpectrum, had grown to operate cellular systems in 18 markets and have a minority interest in 11 metro areas. The company had almost $25 million in revenue. The following year, however, Centel came knocking. Centel was fighting off a hostile takeover attempt and was willing to pay a large premium for TeleSpectrum as part of creating a "poison pill" to dissuade the acquirer. It was a price that United couldn't refuse. In May 1988 United sold its TeleSpectrum business to Centel for $750 million and United was out of the wireless game almost before it had even started.

But, providentially, as we mentioned earlier, Sprint acquired Centel in 1993 and was back in the wireless game. With the merger Sprint had cellular operations in 42 metro markets, had an equity interest in 31 other metro markets and 79 rural markets covering a total combined proportional population of over 20 million people. Sprint Cellular continued to grow surpassing 1 million subscribers in 1994 and $834 million in revenue in 1995.

In evaluating external trends, the company recognized the significant growth potential in wireless, and the growing importance of data communications. Looking inwardly, strategy leaders recognized that the local telephone business was stable and profitable, but not growing, and the long distance business was still growing, but approaching its peak. It was time for another big bet, and that bet was in wireless.

The first generation of cellular service was analog. The second generation was digital but very limited. The FCC had announced an auction for new spectrum bands which could deliver higher data speeds and that would support Personal Communications Service (PCS), supporting voice, data, personalized content, pictures, and eventually video. This auction provided the opportunity for the company to build an "all-digital, nationwide network" that would complement its fiber network and voice, data, and video wireline services.

As Sprint's 1995 Annual Report put it "Sprint Spectrum will create and market a higher quality and more reliable digital wireless service than is known in today's marketplace. Using the most modern technology on a consistent, national basis to achieve a distinct competitive advantage may well be a rerun of how Sprint redefined the long distance market by building the nation's first, and still only, all-digital fiber-optic network. Sprint Spectrum has already acquired licenses to provide the next generation of wireless service known as Personal Communications Service (PCS). We will package this new PCS wireless service with our long distance and local service to create a comprehensive offering that will meet all the needs of our customers from a single, well-respected source. Sprint Spectrum's reach will cover more than 182 million people, nearly three-quarters of the U.S. population, giving us the greatest coverage of any wireless provider in the United States. We will offer better clarity, more privacy and greater value than existing wireless service."

By the time that Annual Report was released, the company had taken two critical steps in this big bet on wireless. First, they formed a joint venture (Sprint Spectrum LP) with three of the largest cable providers in the country: TCI, Comcast, and Cox. That partnership then aggressively participated in the FCC's PCS auctions. Together they spent $2.2 billion to acquire the 29 licenses to provide the coverage mentioned in the annual report. The company would spend another $10 billion dollars building out the network to put that spectrum to work, and another $2 billion (without partners) for more spectrum to fill in the nationwide footprint.

One cost of this big bet was that FCC rules forced Sprint to divest its existing wireless operations. In early 1996, Sprint Cellular was spun-off to Sprint shareholders as an independent entity. It was re-named 360 Communications. In 1998 Alltel acquired 360 for $5.8 billion. Ten years later Verizon Wireless acquired Alltel for $28 billion.

By the end of 1996 Sprint PCS offered service in 8 markets. By the end of 1997 that was up to 134 metropolitan markets, and by the end of 1998 the network had expanded to cover more than half of the U.S. population. While AT&T had once predicted that the total number of wireless subscribers nationwide would top out around one million, Sprint PCS added 836,000 new customers in the fourth quarter of 1998 and had reached the 3 million subscriber mark by February of 1999. In November 1998 Sprint recapitalized its common stock, creating a new PCS tracking stock that was used to buy out the cable partners. Just like with long distance, Sprint had leveraged partners to share startup costs and fuel startup growth, and then taken control of the company's new growth engine when its partners couldn't continue to support the steep losses required to fund that growth.

In his letter in the 1998 Annual Report, chairman Bill Esrey wrote "Sprint PCS is the only wireless service using one brand, one network and one digital technology on a nationwide basis. ... By 2007, total annual revenues for the industry are expected to approach $90 billion. We are uniquely positioned to earn a significant share of this opportunity."

Esrey and team were early in recognizing that wireless is different from traditional telephone service in one important way — the telephone user moves. Throughout the history of the industry, someone could build a successful telco business by serving a very narrow geography. But with wireless, a nationwide footprint became critical for success. Customers wanted their phone to work wherever they went and they didn't want to be surprised by roaming charges on their bill at the end of the month.

Sprint also continued to be a leader in technology innovation being early to market with cameraphones[1], picture mail[2] services, and smartphones[3].

---

1. https://www.ign.com/articles/2003/05/06/sanyo-scp-5300-review

Wireless quickly became the growth engine for the company. In 2000 Sprint PCS reached $6.3 billion in revenues (but with an almost $2 billion operating loss), surpassing the Local division as the second largest part of the company. Growth continued and in 2002, with over $12 billion in revenues, Sprint PCS surged past the Global Markets (long distance) division. That year the wireless division also finally achieved profitability. By 2004 Sprint PCS was larger than the local and long distance businesses combined.

Sprint had become a wireless company.

---

2. https://newsroom.sprint.com/
sprint-allows-family-and-friends-to-share-photos-with-military-unable-to-be-home-this-holiday-se
ason.htm

3. https://www.ign.com/articles/2002/03/19/handspring-sprint-treo-service

One Sprint

─────

T he last three chapters have discussed the origins of the three main divisions of Sprint heading into the 21st century. For most of the company's history, United Telecom was primarily a local telephone company. In the 1980s, the company expanded into long distance, even renaming the entire company Sprint, around that long distance brand. The final piece came into place in the 1990s with Sprint's big bet on PCS wireless. On October 13, 1998, before Sprint had even taken full control of PCS, Sprint president Ron LeMay announced the formation of a new "One-Sprint" organization to pull the pieces together into value-creating offers for customers.

As with most of Sprint's strategic moves, this was a first step in a long road to a new reality.

Before exploring Sprint's journey to integrating local, long distance, and wireless, it is worth stopping and considering how these three components fit into a strategic portfolio.

Business lifecycles are often described using the Sigmoid mathematical function as shown below. We call this an S-curve.

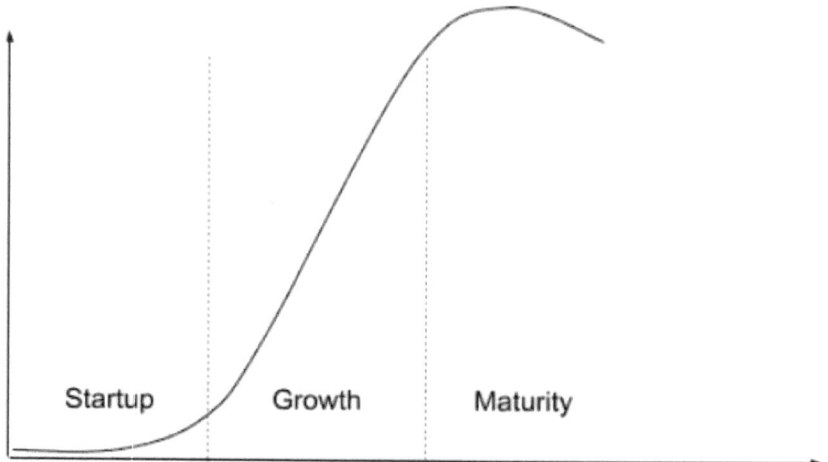

In the diagram, I've marked three phases: Startup, Growth, and Maturity. In each of these phases, your strategy changes. What you want out of the business is different in each phase. What it takes to continue being successful is different. Recognizing the transition from one phase to another is really important so that you can shift your strategy as appropriate.

In the late 1990s the local division was clearly a mature business, long distance was nearing the end of the growth phase, and wireless was a startup business.

Businesses often talk about "jumping to the next S-Curve" — meaning that, before the company's core product or market reaches end of life, they want to have transitioned to a new product or market which is early in the growth phase (see below). Companies will continue to operate the old business to provide the cash being invested into the new business through its startup phase, but over time the focus shifts from the old to the new.

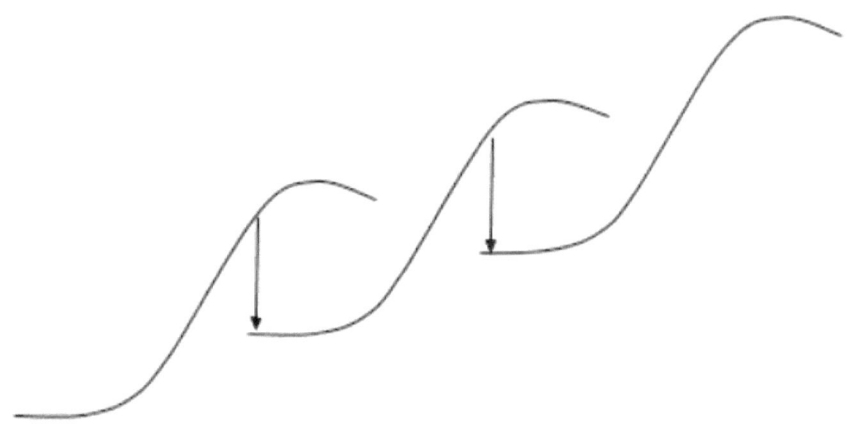

Into the late 1990s, Sprint had managed these businesses masterfully. The company had managed to jump to the next S-curve twice, once in the transition to long distance and then to wireless. This is one of the most difficult strategic moves any company can attempt, and Sprint had managed to pull if off, not once, but twice. The company was managing its portfolio of businesses well and all stakeholders were reaping the benefits.

But now, starting with One Sprint, the company started down a path that would become much more difficult.

I joined Sprint in 2003 as director of strategic planning for GMG, the long distance division. The corporate strategy organization did a good job of coordinating activities between the three different divisions of the company so I quickly became familiar with both the local and wireless divisions as well. The differences in culture and resulting strategies were stark.

The local division was a mature, highly regulated utility business. It had long been the source of cash for investments in Sprint's new ventures. Management focused on operational excellence to maximize returns on any investments made into the business. The status quo had been very good for this business and so management fought hard, internally and externally, to maintain that status quo. Regulations were the local division's best friend.

In late 2003 the long distance division (where I worked) had also recently reached maturity, but much of the organization still acted like it was in growth mode. The consumer long distance base still provided good income, but most of the focus had shifted to business and government customers. This led to a very sales-driven culture. The customer was "always right" and the organization was focused on delivering solutions that met the needs of each customer.

The wireless division had moved into the growth stage but still very much had a startup culture. As a consumer-focused company, marketing drove the business. Sprint PCS was rightly proud of their innovations. Leaders in the wireless business were constantly challenging the status quo and regulations were an inhibitor to innovation and growth.

Early in 2003 Sprint had decided to take One Sprint to its full extent by reorganizing the company. The old product-defined divisions would be replaced with new customer-defined divisions. Due to regulatory requirements, one of the new divisions was the consumer portion of the local telecom division. The other two new divisions were a business customer-focused division and a consumer-focused division. Naturally, the core of these two divisions were the old long distance and wireless divisions respectively. This reorganization was called the Transformation.

As you can imagine, given the cultural differences, this would not be an easy transition.

In October of 2003, I got an email from Howard Janzen. We'd worked together in the past and Howard had recently taken the job as president of Sprint's long distance division. Howard is smart and steady as a rock, even in tough times, which is probably a big part of why he got the job.

By this time, the entire long distance industry was in serious decline.

The Internet emerged on consumers' radar in the late 1990s to much hype. Investors were irrationally exuberant about every new Internet business model and the long distance companies jumped on the bandwagon. When that bubble burst in 2000, many telecom companies went bankrupt. Others had to write down significant Internet investments. And of course, the entire economy went into recession.

During the recession, people looked to cut costs, including long distance calling. Meanwhile, they had new tools at their fingertips to be able to stay connected even without making that expensive long distance call. Many had started using email and nearly half the population had started to use cellphones, which typically didn't charge extra for long distance calls.

By 2003, the economy was recovering, but long distance calling wasn't. For the first time ever, the total number of minutes people called long distance declined, falling 5% in just two years to less than 800 billion, and the average price per minute had dropped from almost 10 cents to less than 8 cents (a net revenue loss of over $19 billion). On top of that, the Telecom Act of 1996 had allowed the monopoly local Bell companies to start offering long distance after they completed a number of pre-conditions. From 2001 to 2003, the local companies had grown their long distance revenues from $6.3 billion in 2001 to $8.7 billion in 2003, taking another $2.4 billion in already shrinking revenue away from traditional long distance companies like Sprint.

Then Sprint decide to start the Transformation. Thus, Howard's email. He was changing jobs to be the president of Sprint Business Solutions and would likely spend the next few years ironing out the strategy. Was I game?

Within a month, I was on-board and working with Howard's team to begin the process. Over the next few years, we would work through a series of critical strategic decisions, but we had an immediate need for a strategic framework. There was a Sprint Board meeting in early December. A large management consulting team had done great analysis of the new business, but what Howard needed was the compelling story. What was this business all about and how was it going to happen?

In less than two months we had a story and a start to a strategy. The overarching mission was to "destroy legacy industry barriers that hinder the success of our business customers." The rallying cry version of the same was, "We Make Business Work!" The three pillars supporting the strategy were:

- Deepen relationships with customers to grow revenue and profits
- Be the leader in wireless/wireline integrated solutions
- Be the easiest to do business with

For each of these pillars we identified three strategic priorities and developed seven strategic programs across the three pillars to help implement the priorities. We also identified 32 key metrics to track progress toward executing the strategy.

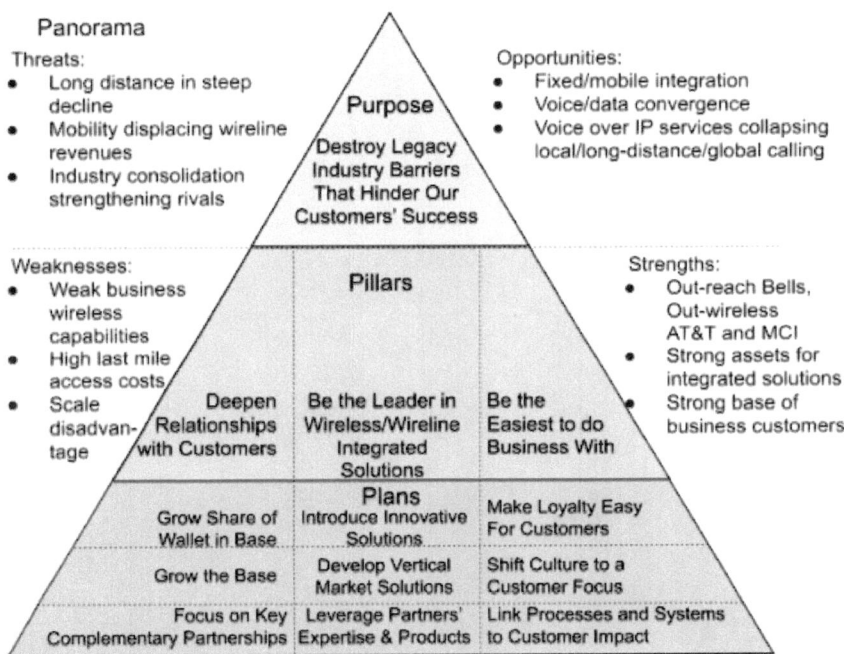

Panorama

Threats:
- Long distance in steep decline
- Mobility displacing wireline revenues
- Industry consolidation strengthening rivals

Purpose

Destroy Legacy Industry Barriers That Hinder Our Customers' Success

Opportunities:
- Fixed/mobile integration
- Voice/data convergence
- Voice over IP services collapsing local/long-distance/global calling

Weaknesses:
- Weak business wireless capabilities
- High last mile access costs
- Scale disadvantage

Pillars

Strengths:
- Out-reach Bells, Out-wireless AT&T and MCI
- Strong assets for integrated solutions
- Strong base of business customers

Deepen Relationships with Customers | Be the Leader in Wireless/Wireline Integrated Solutions | Be the Easiest to do Business With

Plans

Grow Share of Wallet in Base | Introduce Innovative Solutions | Make Loyalty Easy For Customers

Grow the Base | Develop Vertical Market Solutions | Shift Culture to a Customer Focus

Focus on Key Complementary Partnerships | Leverage Partners' Expertise & Products | Link Processes and Systems to Customer Impact

Over the coming months, the transformation happened. Formerly Sprint long distance, wireless, and local employees suddenly found themselves working for a new business. Howard and his team stood in front of these employees with a clear and compelling plan for how they would win in the marketplace. That strategy gave employees a vision for the future and helped them make day-to-day decisions.

Was it a perfect strategy? No. But it set the right direction.

There were still many hard, strategic decisions to be made. We still needed to figure out how to stabilize the long distance business. We needed to work through which customers fit the new strategy and how to attract new prospects. The product pipeline needed to be developed and refined. And we needed to significantly strengthen our wireless solutions capabilities. We would work through each of those in the coming years.

But for now, we could start moving in the right direction.

Sprinting to Nextel

I remember Steve, my boss at Sprint, coming into my office in mid-October 2004 with a grave look on his face. It had been almost one year since I joined the company as director of strategic planning for the business solutions division.

The two of us had immediately hit it off. We had very complementary strengths, similar values, and both were loyal to the company. While Steve's responsibilities were broader than just strategy, we worked well together and had an ongoing open dialog about everything happening in the company, good and bad, so that we could help the president of our division manage potential issues before they became crises.

I asked Steve what was up. "I can't tell you, but it's bad."

"Why can't you tell me?"

"Only a handful of people know and we've been threatened with termination if we tell anyone."

"It", we would all later learn, was a merger being negotiated between Sprint and Nextel.

Sprint had been having a good run. Between the end of 2002 and when rumors of the merger started to get whispered in October of 2004, the company's stock had risen 38% from $14.48 to just over $20. And that was an especially strong recovery from concerns that Wall Street had in 2003 about strategy and execution that had sent the stock down to $11.

But over the same timeframe Nextel had taken off like a rocket. At the end of 2002 the stock was trading at $11.55. By mid-October 2004 it was above $25 — an increase of 120%! Nextel had built a solid and very profitable business by focusing on business mobile customers — especially blue collar workers who loved the company's push-to-talk (PTT) feature. The Nextel salesforce sold businesses on the value of getting more work done, which made it easy for small business owners to justify buying cellphones for workers, even if they didn't know how to calculate an ROI (return on investment).

The average monthly bill for a Nextel customer was nearly $70 compared to the industry average of $52. That was attractive, but what sent the stock price flying was Nextel's aggressive expansion into the consumer market — opening retail stores and launching a national advertising campaign. Nextel grew its consumer subscriber base from about 3 million at the end of 2002 to about 5.5 million by the end of 2004. Wall Street was sold on the potential for Nextel to maintain high margins while expanding into a much larger market.

On the surface, this looked like a perfect marriage. Sprint already had a strong consumer presence (15 million subscribers at the end of 2004), but lacked much of a presence in the business wireless space (2.5 million subscribers compared to Nextel's 10.5 million). Sprint did sell into businesses, especially medium to very large businesses, but mostly long distance voice and data products, not wireless. Combining Nextel's business wireless offers and their strength in the small to mid-sized business segment with Sprint's strength in consumer wireless and the mid-to-large business segment looked promising.

So, why was Steve so concerned?

His first concern was that Sprint was going to grossly overpay for Nextel. The transaction was going to be a merger of equals, even though Sprint had more wireless subscribers, more wireless revenues, and had significant local telco and long distance businesses making Sprint more than twice as big as Nextel. But because Wall Street was overly confident in Nextel's expansion plans, the two companies were being valued equally in the transaction.

Nextel's core business was very profitable, but its potential to grow was much more constrained than Wall Street recognized. The company had no meaningful data strategy, and the future of wireless was data (just look at your smartphone). Their business was operating on a 2G network (about 100 kbps for data — the speed of dialup modems). They didn't have the ability to build a 3G network (about 1 Mbps — the speed of telephone line-based digital subscriber line or DSL broadband), and only had 4G spectrum in about ⅓ of the country. Sprint was a leader in data services with a nationwide 3G network. Sprint also had 4G spectrum in about ⅓ of the country. Without Sprint's data capabilities, Nextel would not be able to participate in the smartphone market. Also, Nextel's push-to-talk was wildly popular with field workers, but would never become popular with housewives or business executives.

But Steve's real concern was that the combining companies lacked a strategy.

The industry had been consolidating and both companies felt an urgency to become bigger. At the beginning of 2003, there were 13 wireless providers with at least a million subscribers. Verizon was the largest with 34 million subscribers (22% share). In order, the next 5 were AT&T Wireless (14%), Cingular (14%), Sprint (11%), T-Mobile (8%), and Nextel (8%).

By 2004, Cingular and AT&T had combined to jump to nearly 47 million subscribers, and the two Bell companies (Cingular and Verizon) were running away from the pack with a combined 49% market share.

Verizon was rumored to be eyeing acquiring either Sprint or Nextel to retake the lead and both companies wanted to control their own destinies. Each knew that if Verizon bought the other, they would be at that much more of a scale disadvantage, without an obvious way to catch up. By combining, Sprint and Nextel could escape Verizon's clutches and start to close the scale gap (combined 19% market share).

But "grow at any cost" is rarely a winning strategy. In fact, it's no strategy at all.

Hindsight is almost always 20/20, and many have been critical of the Sprint Nextel merger, but Steve saw this looming stragedy and tried to warn his superiors, but the course was set.

What is a Stragedy?

ONE TIME MY TEAM RECEIVED an e-mail from a co-worker where he misspelled strategy as stragedy. Unfortunately, since many of the things called "strategy" really aren't, they often do lead to "tragedy," perhaps making my co-worker's spelling justified (at least for those so called "strategies"). For many years, I used the phrase with my strategy compatriots as insider talk when working on projects that were challenged by executives' mistaken belief they had a strategy. We didn't know it had gained wider use.

The Urban Dictionary defines a stragedy as "a plan that has the conviction of forethought, but which is so disastrous it begs for hindsight before it is even implemented. A stragedy is often promoted as a strategy to hide the fact that it was conceived from rash and not rational action."

The Sprint Nextel merger fits the definition.

The Sprint Nextel Tragedy

MANY[1] believe[2] that[3] the[4] Sprint[5] Nextel[6] merger[7] is[8] one[9] of[10] the[11] biggest[12] failures[13] in[14] business[15] history[16]. Many factors contributed to the disaster.

The companies' wireless networks used competing technologies. It would take 10 years before the combined company would be able to standardize on a single technology and replace all of the incompatible cellphones in use by long time customers. In the meantime, the company had to continue to invest in capacity and coverage for two radically different platforms, continue to operate both, and continue to work with partners to develop customer devices to run on both of them.

---

1. https://www.investopedia.com/articles/financial-theory/08/merger-acquisition-disasters.asp

2. https://www.washingtonpost.com/wp-dyn/content/article/2007/11/23/
   AR2007112301588.html

3. https://blogs.wsj.com/deals/2008/02/28/sprint-nextel-officially-a-deal-from-hell/

4. https://www.ft.com/content/e8e2686e-765e-11dc-ad83-0000779fd2ac

5. https://www.axial.net/forum/4-deals-that-failed-and-why/

6. https://www.businessinsider.com/2008/2/sprint-nextel-q4-earnings-analysis?op=1

7. https://tweakyourbiz.com/management/sprints-disastrous-mistake-can-learn

8. https://www.fiercewireless.com/wireless/
   report-sprint-s-merger-nextel-ranked-among-least-successful

9. https://www.cnbc.com/2009/12/29/Top-10-Best-(and-Worst)-Mergers-of-All-Time.html

10. https://www.wsj.com/articles/SB116048367524088013

11. https://teamtophat.blogspot.com/2011/03/culture-clash-sprint-nextel.html

12. https://www.cnet.com/news/bloomberg-ranks-sprint-nextel-deal-among-worst-mergers/

13. http://globalbizresearch.org/IAR16_Vietnam_Conference_2016_Aug/docs/doc/PDF/
    VSL614.pdf

14. https://blogs.wsj.com/deals/2007/10/09/sprint-nextel-anatomy-of-a-failed-merger/

15. https://www.workhuman.com/resources/globoforce-blog/
    6-big-mergers-that-were-killed-by-culture-and-how-to-stop-it-from-killing-yours

16. https://www.cnet.com/news/sprint-gets-the-nextel-monkey-off-its-back/

The two companies had very different cultures. Nextel was an east coast startup — aggressive and embracing risk. Sprint was a 100 year old mid-western corporation with a laid back corporate campus. In most combinations, the acquiring company culture wins. As a merger of equals, the combined company kept both headquarters and equally mixed the management team creating pockets of loyalty to legacy brands, legacy networks, legacy products, legacy leaders, and legacy ways of working. Decision-making slowed and when decisions *were* made, they often were undermined by slowness to act or lack of cooperation.

The shareholders of the two companies had different expectations. Sprint had long offered shareholders a dividend. Nextel never had, with shareholders earning their returns thanks to the rapid growth of the company.

	Sprint	Nextel	Alignment
Market Strength	Consumer	Business	Complementary
Core Technology	CDMA	iDEN	Incompatible
Culture	Conservative	Aggressive	Mismatch
Investor Expectations	Dividends	Growth	Mismatch

The companies shared with investors their expectations for cost saving synergies that would be achieved through the merger, and tremendous focus was placed on achieving those synergies. However, without a clear strategy, the true value of the merger could never have been realized.

A solid strategy:

- Is a framework with depth and dimensionality to deal with uncertainty
- Provides long-term vision
- Defines the goal being pursued
- Sets a direction forward

- Identifies critical near-term objectives
- Enables decision-making

The "strategy" (or rather stragedy) behind the Sprint Nextel merger was simply the goal to "get bigger." There wasn't a long-term vision for the combined company. There wasn't a clear direction forward. Near-term objectives were entirely focused on achieving financial synergies. All decisions were hard, especially given the cultural challenges.

Sadly, this was the beginning of the end for Sprint.

Surviving the Sprint Nextel Stragedy

───

M oving beyond the Sprint Nextel merger, I want to focus on four strategies that were critical to the company's ability to survive the Sprint Nextel merger stragedy:

- Long Distance (Wireline)
- Local (Wireline)
- 4G (Wireless)
- Prepaid (Wireless)

BCG Growth Share Matrix

BEFORE LOOKING AT THESE strategies, I want to introduce a tool often used by corporate strategy groups in managing portfolios of businesses or markets.

In 1970, Bruce Henderson of Boston Consulting Group (BCG) developed the Growth-Share Matrix as a way of evaluating the relative attractiveness of different businesses, products, or markets within a corporation.

The tool plots the different businesses against two distinct measures — relative market share (your market share divided by your biggest competitor's market share) and market growth rate (see chart below). The relative size of each business (typically measured in gross sales) is represented by the size of the bubble for that business on the chart. The two axes typically cross at or near the median values of each other, creating four quadrants. The four quadrants have been given names to reflect the attractiveness of businesses in each.

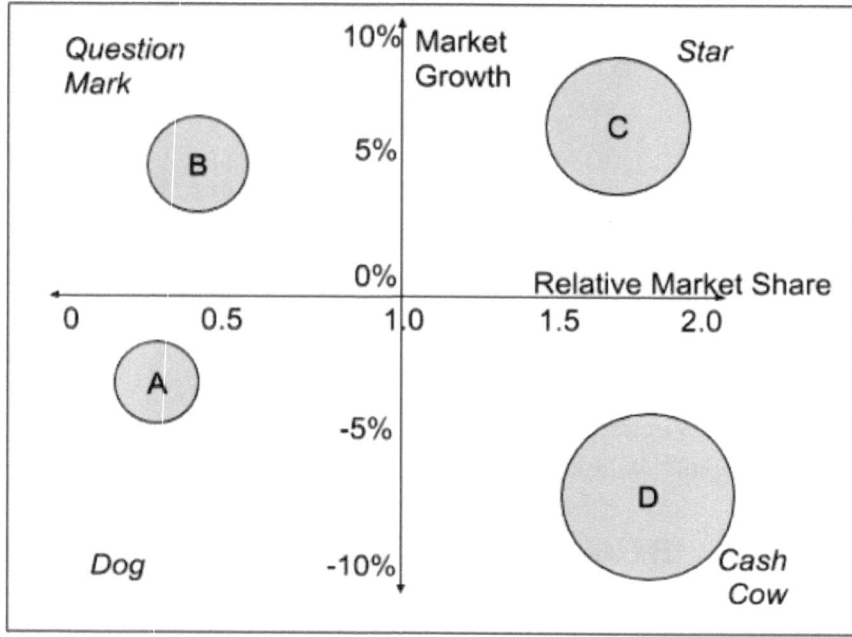

A business that has low market share and is in a market that is not growing as fast as others in which the corporation participates (Business A in the diagram) is called a "Dog."

One that has low share, but in a growing market (Business B) is called a "Question Mark."

A business with high share in a high growth market (Business C) is called a "Star."

One with high share, but in a low growth market (Business D) is called a "Cash Cow."

Corporations will often maintain a portfolio of businesses that are spread across the four quadrants, but the amount and nature of investments can be shaped by understanding the implications of each. Both Cash Cows and Stars have high market share and therefore can be very profitable.

Most corporations will choose to invest in both Stars and Cash Cows, but with different approaches.

Investments in Cash Cows are likely aimed at retaining revenues and share while maximizing margins, while investments in Stars are likely aimed at maximizing growth with attractive margins. A Question Mark is so named because it has the potential to either become a Star or a Dog. Leaders must evaluate the potential to grow share with focused investment, sometimes through mergers and acquisitions. Dogs are the most challenging and often are considered for divestiture or shut down.

BCG analysis can inform the corporate strategy development process, but it shouldn't dictate the outcome. While the Growth-Share Matrix implies internal strengths and external opportunities from which flow financial performance expectations, many other factors come into play. Synergies often exist across business units, making some dogs attractive within the overall portfolio. Sometimes the strengths that have enabled a Star to achieve high share are not the same strengths required to maintain leadership as the market grows and matures.

Long Distance

The first of the four key strategies actually preceded the merger with Nextel.

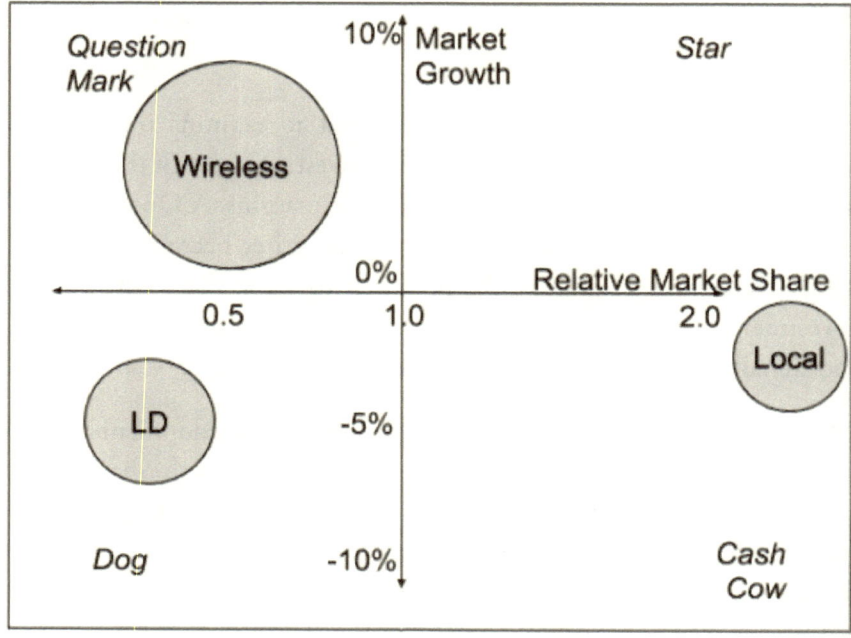

If we look at the BCG Matrix for Sprint's three traditional businesses at the time, we see that Wireless was a Question Mark (about half the size of Verizon, with approximately 4% industry growth), Long Distance was a Dog (about one-third the size of AT&T, but with -5% industry growth), and Local was a Cash Cow (a near monopoly in each of its markets, but with -1% industry growth).

The merger with Nextel was intended to position the Wireless business to move from Question Mark into the Star quadrant.

The big challenge was the Long Distance wireline business. Being in the Dog quadrant is bad enough, but worse, while the business had produced over a billion dollars in cash per year in previous years, that performance was declining rapidly and was expected to go cash flow negative within a few years. People simply were disconnecting their physical telephone lines and going mobile, and that was not a trend likely ever to reverse itself.

Strategic options were considered at the corporate level ranging from exiting the wireline business, to "doubling up" by acquiring AT&T (later acquired by SBC) or MCI (later acquired by Verizon). Exiting the business was rejected because of the reliance of the wireless businesses on the wireline network. Doubling up was rejected because the overall market was in decline and adding assets wouldn't significantly improve the synergies with the core wireless businesses. Therefore, the "problem" was passed down to the business unit managing the wireline business.

At the time I was still the director of strategic planning for Sprint Business Solutions (SBS), the business unit managing the wireline business. We engaged the SBS management team in an intensive process to identify and evaluate strategic options that would accomplish the role required by the corporate strategy while delivering acceptable financial performance. We considered a range of options, but since the market was shrinking, not growing, most options required significant cost cutting. Building enough detail into the plans for realistic analysis and comparison required working down into the organization and across partner organizations (e.g., IT and Network). Given that the options being considered would likely involve painful decisions around reducing headcount, exiting markets, and eliminating products, all of this had to be done with tight confidentially, both internally and externally.

In the end, we selected and refined a strategy we called, "Extreme Discipline." This strategy included a product strategy to exit traditional "circuit" products (private line, frame relay, ATM) that were being replaced by "packet" (Internet protocol and Ethernet based) products. It also required market strategies for small, medium, and large business markets to focus on where we could win (specific cities for small and medium businesses where we had competitive network assets and specific industry vertical markets for large businesses where we had competitive solutions). This involved selling deselected product capabilities/customers to other companies, shutting down sales offices, reducing sales headcount significantly in deselected markets, and more modestly reducing costs and headcount in all other organizations.

In addition to achieving the targeted financial goals, some of the cost savings were reinvested in growing the wireless business. While the decisions were hard and the strategy wasn't "exciting," the decisions made and executed positioned the company well for the merger of Sprint and Nextel. The cost reductions and investments in wireless growth contributed to overall corporate performance improvements that helped Sprint's stock price to increase from $17 in mid-2004 to $24 in early-2005. Meanwhile, the increased focus on wireless sales in the small and medium business markets aligned perfectly with Nextel's strength in those markets.

Clearly defining the strategy helped communicate hard decisions internally and externally and helped focus execution on critical elements.

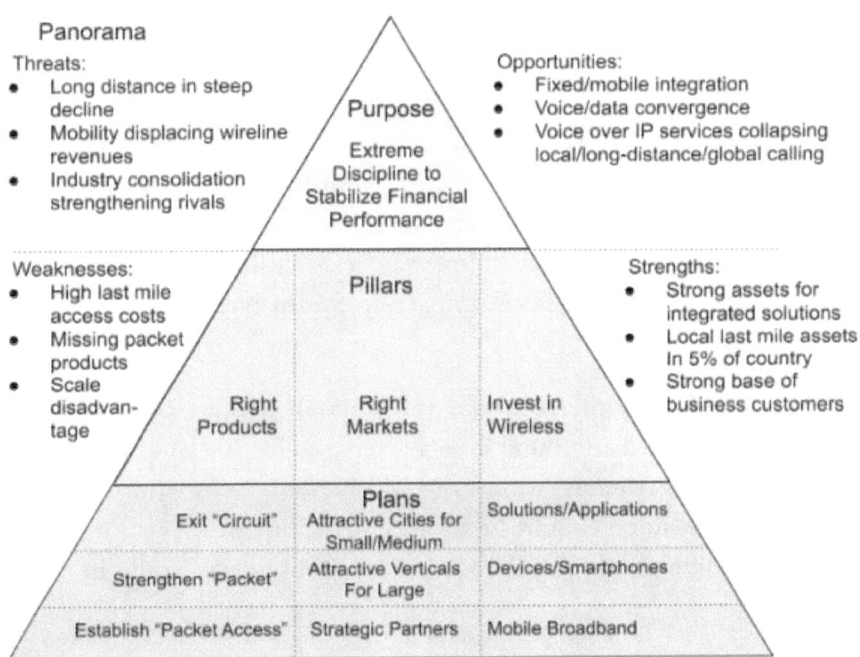

Panorama

Threats:
- Long distance in steep decline
- Mobility displacing wireline revenues
- Industry consolidation strengthening rivals

Opportunities:
- Fixed/mobile integration
- Voice/data convergence
- Voice over IP services collapsing local/long-distance/global calling

Purpose
Extreme Discipline to Stabilize Financial Performance

Weaknesses:
- High last mile access costs
- Missing packet products
- Scale disadvantage

Strengths:
- Strong assets for integrated solutions
- Local last mile assets In 5% of country
- Strong base of business customers

Pillars

Right Products	Right Markets	Invest in Wireless

Plans

Exit "Circuit"	Attractive Cities for Small/Medium	Solutions/Applications
Strengthen "Packet"	Attractive Verticals For Large	Devices/Smartphones
Establish "Packet Access"	Strategic Partners	Mobile Broadband

Although "Extreme Discipline" set our long distance wireline strategy going forward, this topic became a perennial one — every year we would ask the question "why are we still in this business?" Every year the same answer came back — as long as it can support itself, we need the wireline network to provide owner's economics for wireless backhaul; we need long distance voice and data products to provide a complete portfolio and integrated solutions to business customers; and we need the complementary network capabilities for working with strategic partners including cable providers and Clearwire (see section below on 4G).

Local

THE SECOND STRATEGY immediately flowed from the merger of Sprint and Nextel. I don't fully understand all the reasons for doing so, but as part of the merger announcement between Sprint and Nextel, it was also announced that the combined company planned to spin Sprint's local business out as a separate entity.

It is possible that doing so would make it easier to gain regulatory approval. Often large mergers require divestitures to reduce concerns that the combined business will have too much influence on the market, therefore hurting consumers. However, it is unlikely that the assets spun-off would have significant impact on Sprint Nextel's performance in the wireless markets.

It's more likely that the spin was intended to address concerns that investors couldn't clearly value such a diverse collection of assets. Either the high growth wireless business would be undervalued, or the high margin local business would be undervalued, or both. Separating the two would increase the likelihood that both businesses would be fairly valued.

But perhaps one of the biggest drivers for the spin was the opportunity to transfer much of Sprint's long term debt to the new local business, leaving the combined Sprint Nextel with (what at least seemed to be) a reasonable debt load. At the end of 2006, the new local business reported $6.4 billion in long term debt.

As mentioned in the previous chapter, there were significant cultural mismatches between Sprint and Nextel. Nextel acted as a very aggressive East Coast tech startup, while Sprint was more conservative. Within Sprint, the local division was even more conservative than the wireless and long distance divisions, especially on regulatory issues, so spinning off the local division would, to some small degree, reduce these cultural challenges.

Similarly, I mentioned that there were differences in investor expectations. Sprint paid dividends while Nextel invested generated cash back into growing the business. The local spin-off announcement indicated that the new local business would continue to pay dividends and implied that the combined Sprint Nextel would not. So, perhaps the spin was also intended to satisfy both sets of investors.

The new local business also provided a soft landing spot (for a time) for a number of Sprint executives who were not selected for the combined Sprint Nextel. It's possible that retaining jobs and talent in the Kansas City telecom industry was a contributing factor to the decision.

Whatever the reasons were, the Sprint Nextel merger was completed in August 2005 and the local business was spun off as an independent business named Embarq in May 2006. This was not a small company. Embarq was the largest independent telephone company in the country, serving customers in 18 states, generating approximately $6 billion a year in revenue, and employing approximately 18,000 workers.

However, as the traditional telco industry continued to be pressured by revenue declines and new competitors (most notably cable companies), consolidation continued. On October 27, 2008, Embarq announced that it had agreed to be acquired by CenturyTel. That company, now called CenturyLink, has continued to consolidate the industry and is by far the largest independent telco in the country.

4G

THE THIRD STRATEGY didn't so much flow as stumble out of the merger. As I mentioned in the previous chapter, Sprint had 4G spectrum in about one-third of the country and Nextel had 4G spectrum in about one-third of the country. Combined, that still didn't provide a nationwide footprint for building out a 4G network. Even so, the company created a new division, named Xohm, to pursue the buildout of the new network. That network would be the first 4G network in the country.

There was a lot of work to do. The spectrum that both companies brought to the table was in the 2.5 GHz Broadband Radio Spectrum (BRS) and Educational Band Spectrum (EBS) bands. These bands supported much larger channels than had previously been used for wireless data, meaning that end users would receive more bandwidth and would experience faster performance.

However, the 2.5 GHz spectrum was not yet part of the international technology standards and hadn't yet been broadly adopted around the globe. This spectrum also lent itself to a different approach to downlink and uplink engineering than had traditionally been used in the industry. Cellular networks usually use Frequency Division Duplexing (FDD) meaning that spectrum comes in paired channels with half the spectrum being used for uplink from the cellphone to the tower and the other half being used for downlink from the tower to the phone.The BRS/EBS bands weren't paired and instead used one band for both uplink and downlink with Time Division Duplexing (TDD), meaning the uplink and downlink took turns using the spectrum. TDD can be much more efficient than FDD, which is good, but it was another non-standard aspect of Xohm's spectrum.

All of that meant that equipment manufacturers had to be convinced to build cellphones, data cards, and network equipment for this spectrum. And since the cost of these devices and equipment only comes down with high volumes of manufacturing and sale, Xohm had to convince other wireless operators around the world to adopt this spectrum band and the TDD approach. Xohm also had to actively work with all of the standards bodies to get the BRS/EBS bands included in standards, and to work with regulatory bodies around the world to help other operators commit to the 2.5 GHz bands. It was a lot of work and took many months of time and global travel to build the momentum to launch the business with confidence.

Meanwhile, Xohm also had to choose a networking standard. There were two main candidates. LTE (long term evolution) was the natural evolution from the cellular networks deployed around the world and had the strongest support within the traditional telecom community. However, WiMax was gaining momentum especially in technology industries. Intel was a big supporter of WiMax, and WiMax networks were already being launched around the world, while LTE was still years away. Most notably, Clearwire, a startup founded by wireless industry legend Craig McCaw, had committed to WiMax. Clearwire also controlled the 2.5 GHz spectrum in the one-third of the country not covered by Xohm.

While all of this hard work was happening at Xohm, things weren't going well for the new Sprint Nextel.

Everything looked pretty good at the beginning. At the end of 2005, the company had nearly $9B in cash. In the first quarter of 2006, the company added 563 thousand postpaid wireless subscribers, had industry leading ARPU (average revenue per user) of $62 per month, and reported net income of almost half a billion dollars.

But over the next two years, the business turned around, in a bad way. The company's two networks were having performance challenges probably due to splitting network investment and attention between the two. Customer service was also struggling. Customers lost confidence in the company and the brand suffered. Worst of all, because the company lacked a competitive strategy, there was no compelling reason for anyone to choose Sprint over its rivals. In the fourth quarter of 2007, the company wrote down $29.7 billion in goodwill, reflecting that almost all of the value of the merger had been lost. It was one of the largest write-downs[1] in corporate history. At the end of 2007, the company's cash was down to $2.2 billion (with $20 billion in long term debt).

---

1. https://seekingalpha.com/article/1561532-asset-write-downs

In the first quarter of 2008 the company lost over a million postpaid subscribers, ARPU had fallen to $56, and the company reported a net loss of over half a billion dollars, almost the exact opposite of just two years before.

Tim Donahue, CEO of Nextel and Chairman of Sprint Nextel had left the company in December 2006. Gary Forsee, Chairman of Sprint and CEO of Sprint Nextel followed him out the door in October 2007. In December of 2007, Dan Hesse was hired as Forsee's replacement. Hesse often said[2] that he knew things were bad when he took the job, but he didn't realize how bad they were.

While Xohm had been launched by Donahue and Forsee with great expectations, by 2008 there simply wasn't enough money to actually build it. And so, the strategy had to shift.

By this time I was in corporate strategy, and obviously we were still struggling with a number of issues. Keith Cowan was the president over strategic planning and corporate initiatives. Keith is a brilliant dealmaker, and he put together a deal for the ages.

On May 7, 2008 Sprint Nextel announced the combination of Xohm with Clearwire to form the new Clearwire. The new venture received Sprint Nextel's 2.5 GHz spectrum (resulting in a nationwide footprint), and $3.2 billion in cash from Intel, Google, Comcast, Time Warner Cable, and Bright House Networks. The new Clearwire would build and operate the 4G network, and Sprint, the cable companies, and other wholesale customers would sell the capacity on that network to their customers. Sprint owned 51 percent of the new company. So, the strategy was to use other people's money to build the network that Sprint needed to create competitive differentiation.

---

2. https://gigaom.com/2012/12/16/

a-gigaom-conversation-with-sprints-dan-hesse-on-five-harrowing-years-as-ceo/

Clearwire itself would face many challenges along its path, but Sprint was able to claim the first 4G offering among nationwide mobile operators, and when the company launched the first 4G data cards, 4G WiFi hotspots, and the first 4G smartphone, customers finally had a reason to choose Sprint over its competitors. In 2013, Sprint was able to acquire the portion of Clearwire that it didn't already own and take full control of its 4G network.

Prepaid

THE OVERALL TELECOM market continued to be very dynamic, as reflected in the BCG Matrix for the company's businesses as of 2009.

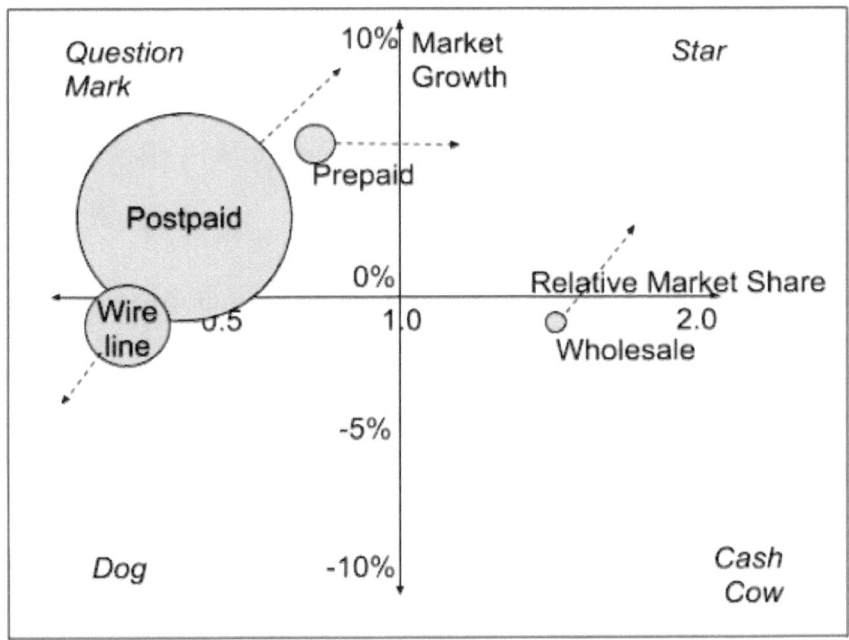

At this point, the company could be evaluated by four key product/market businesses: wireline services, wholesale wireless, prepaid wireless, and postpaid wireless. Wireline services are the traditional "long distance" non-wireless residential and business voice, data, and Internet services. The wholesale wireless business made Sprint Nextel's wireless networks available for resale by other companies under their own brands (e.g. Disney, Kroger). In 2009 prepaid wireless services were primarily wireless voice services provided on lower end phones (not smartphones) for a lower monthly price paid in advance. Postpaid wireless was the standard wireless voice and data services provided on a variety of wireless devices (including smartphones) with the bill paid in arrears. At the time, wireline was (still) a Dog, wholesale wireless was a (very small) Cash Cow, prepaid was a Question Mark, and postpaid wireless was on the border between Question mark and Dog. There were no Stars in the portfolio. (See Figure 5.2.)

There were some emerging market opportunities (primarily around the Internet of Things) that could potentially lift either wholesale or postpaid into high growth (Star or Question Mark respectively), so investments were focused in both businesses to help develop that opportunity.

The big investment decision was to acquire prepaid competitor Virgin Mobile and further invest in prepaid to move it from Question Mark to Star. Sprint announced plans to acquire Virgin in July 2009 and completed the deal in November of that year. Sprint also announced Boost Unlimited, an aggressive prepaid plan which attracted many new customers to the company. The prepaid business became the near-term growth engine for the company. From 2009 to 2010, postpaid revenues declined from $23.2B to $21.9B, but prepaid more than made up for the decline, nearly doubling from $2.1B to $3.8B.

Through this series of strategic actions, the company was able to survive the stragedy of the Sprint Nextel merger. The long distance wireline business was stabilized. The 4G network was built which would provide the competitive differentiation necessary to save the postpaid wireless business. And in the meantime, the prepaid wireless business became, for a time, the growth engine of the company, making up for revenue losses in the rest of the business.

Survival today, however, doesn't guarantee survival tomorrow.

Sprinting Through the Hesse Years

As I mentioned in the last chapter, Dan Hesse officially joined Sprint Nextel as CEO in December 2007.

The merger had closed in August 2005 with great hope and anticipation that the combined company would be able to directly compete with AT&T and Verizon. Remember, the motivation for the merger was simply to get bigger. Telecom is a very asset intensive industry and so scale matters. You need to spend roughly the same amount of money (tens of $billions) to build a nationwide wireless network and open stores everywhere to sell that service whether you have a few customers or a lot of customers. Therefore, you want to have a lot of customers paying their bills every month to provide a return on that up-front investment.

When the merger was announced in December 2004[1], Sprint had about 20.1 million customers and Nextel had about 15.3, so combined they would have 35.4 million. Earlier that year AT&T Wireless had been acquired by Cingular to form the largest provider at 46 million. Verizon was the second largest at 42 million. But the wireless industry was growing rapidly, so by the time the deal closed 9 months later[2], AT&T had 50 million subscribers, Verizon had 45.5 million, and Sprint Nextel had 44 million. The company appeared to be closing in fast on its two larger competitors. On a pro-forma basis, the two companies combined had $40.8 billion in revenues in 2004. Sprint Nextel was officially one of the big boys. (By the way, at the time, T-Mobile had 18.2 customers, so a very distant fourth.)

---

1. https://money.cnn.com/2004/12/15/news/fortune500/sprint_nextel/

2. https://www.networkworld.com/article/2313381/sprint--nextel-close-merger-deal.html

Unfortunately, it didn't stay that way for long.

	Postpaid Net Adds (000s)	Net Operating Revenue ($B)	Net Income ($B)	Ending Stock Price	Long-Term Debt ($B)
2006	-279	$41.0	$1.3	$18.89	$21.0
2007	-1,224	$40.1	-$29.6	$13.13	$20.5
2008	-4,073	$35.6	-$2.8	$1.83	$21.0
2009	-3,546	$32.3	-$2.4	$3.66	$20.3
2010	-855	$32.6	-$3.5	$4.23	$18.5
2011	-98	$33.7	-$2.9	$2.34	$20.3
2012	-1,137	$35.3	-$4.3	$5.67	$24.0
2013	-2,209	$35.5	-$3.0	$10.75	$32.0
2015	-430	$34.5	-$3.3	$3.85	$32.5
2016	-1,670	$32.2	-$2.0	$8.42	$29.2
2017	811	$33.3	-$1.2	$5.89	$35.9
2018	710	$33.6	-$1.9	$5.84	$35.4

Sprint Nextel Performance, Source: Company Reports (Note, the company's fiscal year changed in 2014–2015)

In my opinion, because the combined company lacked a clear strategy forward, making decisions was hard. The loudest or most persuasive voice in the room won the argument, and those rooms were split between Virginia and Kansas. As executives started to leave, different voices became the loudest. Decisions clashed. The direction forward changed dramatically and often. Money was wasted. Employees were confused. Customers fled.

The chart above shows the long term impact.

From 2006 to 2016 the company lost 15.5 million postpaid subscribers (that's more than the total number of subscribers Nextel originally brought to the merger) and reported $22 billion in net losses (excluding the big write-down in 2007). Revenues fell from $41 billion to $32 billion. This was during a time when the rest of the industry was rapidly growing and our larger competitors were enjoying solid profits.

In my opinion, without the long distance, prepaid, and 4G strategies we discussed in the last chapter, the company would've been in even worse shape.

What Dan Did Right

I REMEMBER DAN'S FIRST day. He held an all hands meeting with hundreds of employees in the Sprint theater and thousands more on video links.

One of his clear messages was that the company needed one headquarters, not two, and that would be in Overland Park, Kansas. The company also needed just one culture, not two.

Most of Dan's career had been at AT&T, but most recently he'd been hired as Embarq's first CEO, so he was familiar with the Sprint culture and its shortcomings. In that first all hands meeting he specifically called an end to "analysis paralysis", the overly conservative approach to decision-making common at Sprint. But he also called an end to "shoot-from-the-hip" decision-making sometimes necessary at high growth startups like Nextel. Over time Dan would define that culture.

This wasn't Dan's last all hands. As his predecessors had done, Dan hosted company wide meetings each quarter after the company's financial results had been released. Early in his tenure, Dan established three priorities for the business:

- Improve the Customer Experience
- Strengthen the Brand
- Generate Cash

In those all hands meetings, naturally, Hesse would speak to the financial performance that the company had just reported, but especially for this audience, the meeting agenda was consistently structured around the three priorities. For each, he would highlight progress — calling out improvement in metrics and external awards. He also regularly gave out internal awards to employees that were aligned with each of the priorities — along with a small cash reward. Through this consistent communication, everyone in the company knew the three priorities and had a tangible picture of what success looked like.

These three priorities were the pillars of Dan's strategy and he sought to translate them into dramatic, impactful actions. On February 28, 2008 Sprint introduced the "Simply Everything" plan, a game changing offer that combined unlimited voice, text, data, and video with value added services including navigation, Sprint TV, and Sprint Music. At $99, the plan was priced higher than most customers were paying, but it provided radical simplicity and price certainty that attracted many to step up to this higher priced plan.

Dan referred to Simply Everything as a "nuke" disrupting the way the industry typically worked. From that point on, throughout his tenure, Dan launched a new "nuke" every quarter. Some nukes were new pricing plans. Some were industry leading devices. Some were new policies or partnerships. All were aligned to Hesse's three priorities.

Dan also obviously was actively involved in the 4G and prepaid strategies we've previously discussed, along with many other important strategic decisions.

What Dan Missed

DAN'S THREE PRIORITIES provided guidance to much of the company, but at times those priorities were in conflict. How do you prioritize the priorities? Which gets scarce resources: a brand initiative, a customer experience initiative, or cost cutting initiatives? Can we spend cash now on brand or experience initiatives that won't generate cash until several quarters from now?

In my opinion, what was missing was an overarching mission or purpose, a "true north" that would unify the three priorities towards a single goal.

I remember a senior leadership meeting early in Dan's tenure when I tried to lead a discussion towards developing a complete strategy, including that top level mission. Bob Brust, the company's new CFO, cut me off quickly. "We don't need a strategy. We have a strategy. It is to survive. Let's not waste time on strategy and instead focus on how to do that."

So, in all fairness, although Hesse couldn't put it down on paper, his time at Sprint was focused on one goal — survival.

That doesn't mean that my team stopped doing our jobs. We worked well with most of Dan's direct report team to develop a strategic framework to provide guidance to the organization, and those efforts were greatly appreciated.

I recently came across the framework we developed in 2011. The industry vision and mission section reads:

We believe that a "mobility revolution" is under way as wireless services and mobility are increasingly becoming integrated with all aspects of how people work and play and how businesses compete and operate.

Sprint will accelerate and benefit from the mobility revolution by:

- Being the "champion of the people" — making it easy for consumers and businesses to embrace the mobility revolution

- Achieving profitable growth by driving wireless (postpaid and prepaid) substitution for legacy telecom products and by driving substitution or business model changes for non-telecom products and services through our wholesale and open enablement

- Uniting the disruptive forces of the mobility revolution against the duopoly intentions of our giant Bell competitors

Softbank to the Rescue

SPRINT MADE GOOD PROGRESS towards improving the customer experience. In 2010, the American Customer Satisfaction Index identified Sprint as the most improved in customer satisfaction over the previous two years across all 47 industries the study covers. That claim continued to hold for the next three years. That really just meant that Sprint went from being in last place, miles behind its competitors in customer satisfaction, to being in the lead pack. It wasn't enough to drive customer decisions. The company continued to lose market share to competitors.

It was even harder to claim gains in brand strengthening and cash generation.

By 2012 the company was in a very difficult spot. The company's 4G leadership had provided one of a very few bright spots, but competitors had largely caught up. Verizon launched its 4G network in December 2010 and its first 4G smartphone in early 2011. AT&T and T-Mobile were marketing their 3G networks as supposedly offering 4G services (or as we called them "faux g") around the same time, and AT&T finally launched its LTE 4G networks in September 2011. (T-Mobile didn't join the party until 2013.)

But, despite its 4G leadership, Sprint didn't really have a 4G network. Remember, due to lack of cash, Sprint had spun its 4G assets and operations out into the Clearwire joint venture. Sprint owned 50+% of the company, but didn't control it. Meanwhile, the cable companies had lost interest in wireless (again, at least for the moment) and so Clearwire's strategy of being the wholesale provider to Sprint, cable providers, and others, had simply defaulted to being Sprint's 4G provider. Being short on cash themselves, Clearwire was struggling to expand the network fast enough for Sprint to remain competitive.

The obvious answer was for Sprint to buy the part of Clearwire it didn't already own, bring it back in house, and invest in the network. But Sprint couldn't afford the purchase price or the network upgrades.

It was a balance sheet problem that needed a deep-pocketed partner to solve. Needless to say, Keith Cowan and his corporate development team were very busy.

The partner they brought to the table was Softbank and its charismatic leader, Masa Son. Masa has been called the Bill Gates of Japan. For one brief instant in time he had been the richest man in the world, and then the Internet bubble burst. Masa made many investments across all kinds of technology industries, but his biggest bets were in Internet, telecom and wireless.

In 2006 Softbank bought Vodaphone Japan, the distant third place wireless provider in Japan. Softbank renamed the company Softbank Mobile and then made complimentary acquisitions, most notably to bring in needed wireless spectrum. Softbank Mobile used the spectrum to create what arguably was the best performing mobile network in Japan. The company also introduced disruptive offers into the market. The net result was that Softbank Mobile quickly closed the gap with competitors and became the second largest provider in Japan with rapidly improving financial performance.

But Japan offered limited growth potential for Softbank and Masa was looking for a similar opportunity in the U.S. On the surface, Sprint seemed to fit the model well.

In October 2012, Softbank announced their intentions to acquire a 70% stake in Sprint. Shareholders would get $12.1 billion and Sprint would get $8.0 billion in cash, part of which would be used to acquire Clearwire. The deal didn't close until July 2013 and the interim months were filled with rival offers from Dish for both Sprint and Clearwire (which somewhat changed the final terms of both deals) and gaining regulatory approvals.

Unlike in some of Softbank's previous deals, Masa allowed Dan Hesse to remain as CEO of Sprint. However, Dan and Masa often clashed over the direction forward for the company. Dan left Sprint in August 2014. As one headline put it, he left the company "the way he found it: needing a turnaround[3]."

3. https://www.cnet.com/news/sprint-ceo-leaves-carrier-the-way-he-found-it-needing-a-turnaround/

Sprinting Into the Arms of T-Mobile

The last chapter ended with Softbank acquiring 70% of Sprint in 2013 and Dan Hesse leaving the company in 2014. Marcello Claure took over as Sprint's CEO. Over the next five years, the company would take at least three shots at merging with T-Mobile. The first attempt, in 2014[1], would have been an acquisition of T-Mobile by Sprint. The second in 2017[2] likely would've been a merger of equals. The third attempt finally resulted in T-Mobile acquiring Sprint earlier this year.

AT&T Fuels T-Mobile's Rise

REMEMBER, WHEN SPRINT and Nextel merged in 2005, the combined company had 44 million subscribers and T-Mobile had 18.2 million. By 2018, Sprint had grown to about 55 million, but T-Mobile had about 76 million wireless subscribers. How did that happen?

The short answer is that AT&T gave T-Mobile the cash and spectrum to build a competitive network and market it to the masses, and T-Mo took full advantage.

On March 20, 2011 AT&T announced its intention to acquire T-Mobile for $39 billion. At the time, T-Mobile was severely challenged. The company was a distant fourth with 33 million wireless subscribers. (AT&T had 95.5 million.) Its network was thin in most of the country and it was the only national carrier without 4G services. Because of its network weakness, T-Mobile focused its marketing and sales activity in

---

1. https://www.wsj.com/articles/sprint-abandoning-pursuit-of-t-mobile-1407279448

2. https://money.cnn.com/2017/11/04/news/companies/sprint-t-mobile-merger-deal/index.html

a relatively small number of markets where it had competitive coverage and performance.

The deal would make AT&T, by far, the largest wireless carrier, leapfrogging Verizon's 101 million subscribers, and leaving Sprint as a distant third, less than half the size of #2 Verizon.

Everyone expected the deal to go through. Obviously, AT&T didn't expect any problems. They agreed to a breakup fee of $3 billion in cash, $1 billion in wireless spectrum, and favorable roaming rates if regulators blocked the deal.

As with any merger of competitors, Sprint basically had three options. They could not fight it and by their silence support its approval. They could fight it and try to get regulators to block the merger. Or they could fight it, but with the real goal being to gain concessions favorable to Sprint (e.g. regulators forcing AT&T to divest customers, spectrum, or other assets that Sprint may be able to purchase at a discount).

Sprint chose to fight the deal and try to get it blocked. Dan Hesse called it[3] the most important decision he made as CEO. On August 31, 2011 the Department of Justice filed suit to block the merger. AT&T and T-Mobile abandoned the deal and AT&T paid the breakup fee to T-Mobile.

T-Mobile took the cash, spectrum, and roaming and built a competitive network. They then hired John Legere to be their new CEO. Legere had been a protege of Dan Hesse's when they both worked at AT&T. Legere took Hesse's "nukes" concept and recast it as a comprehensive strategy that T-Mobile branded the "uncarrier" strategy. The premise — people don't like the way that wireless carriers treat them. T-Mobile broke the industry rules and people loved them for it. Every quarter T-Mobile took

---

3. https://bgr.com/2012/12/17/sprint-ceo-hesse-interview-att-t-mobile-256325/

wireless subscribers away from its competitors, especially Sprint. In 2015 T-Mobile passed Sprint as the third largest mobile operator in the U.S.

The Rule of Three

IN 1976, THE BOSTON Consulting Group's Bruce Henderson published an article titled "The Rule of Three and Four"[4]. In it, he made the argument that, in a stable competitive market, "[a]ny competitor with less than one quarter the share of the largest competitor cannot be an effective competitor." Mathematically, this leads to the conclusion that a stable market cannot support more than 3 effective competitors.

Business Professor Jagdish Sheth further researched[5] a variety of industries to validate and clarify the concept in the late 1990s and early 2000s, publishing *The Rule of Three: Surviving and Thriving in Competitive Markets* in 2002. His work validated the general concept that a mature industry could only support three "full line generalists" but he observed that any industry likely would also have a number of niche product or market specialists.

The niche players could be very profitable by focusing all of their investments and resources on a very narrow segment which they could dominate. Full line generalists (in our context, this would be the nationwide wireless carriers), on the other hand, have to invest broadly to reach and serve the entire market. In our context that means building and operating a nationwide network, opening stores and managing distribution nationally, and running marketing campaigns on a national scale.

The diagram below is roughly based on Sheth's work. What he calls the "Ditch", I usually call "The Valley of Death." Small firms that try to grow out of their successful niche head into the valley where they tend to die.

4. https://www.bcg.com/publications/1976/business-unit-strategy-growth-rule-three-four.aspx

5. https://www.jagsheth.com/geopolitics-globalization/the-rule-of-three-abstract-paper/

Henderson set the cut-off share for a broad generalist to be an effective competitor at 30% while Sheth places it around 10%. I tend to average the two and see 20%, roughly, as the market share required for a national player to survive.

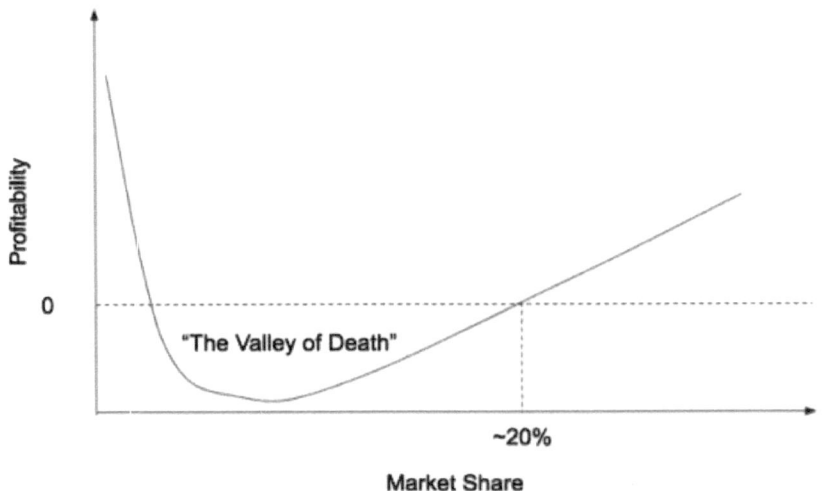

Throughout his tenure, Dan Hesse focused the company entirely on Verizon and AT&T. He wanted the Sprint brand positioned as one of the premium brands and he wanted the employees to be aspiring to the quality represented by those competitors. Prior to AT&T's takeover attempt, T-Mobile had clearly been the "value" or "low cost" brand with an inferior product.

Within corporate strategy, however, we were very concerned about T-Mobile as early as 2008 because of our concerns about the Rule of Three.

Upon completion of the merger in 2005, Sprint had 19% share of subscribers while Verizon and AT&T each had 24%. T-Mobile was far behind at 10%. But by 2008, AT&T had grown share to 27% with

Verizon slightly behind at 25%. Sprint had fallen to 15% and T-Mobile had risen slightly to 12%.

While publicly focused up-market on AT&T and Verizon, we were also working on ways to increase Sprint market share at the expense of T-Mobile to widen the gap and establish Sprint as the sole sustainable #3 player. In addition to the acquisition of Virgin Mobile[6], we launched Boost Unlimited, a disruptive prepaid offer very much in the "uncarrier[7]" mold (long before T-Mobile had formed their uncarrier strategy). The Boost Unlimited offer had a secondary benefit of bringing additional customers and traffic to the Nextel iDEN network which had been rapidly shedding subscribers. We were able to put an underutilized asset to work in attacking T-Mobile's prepaid strength. The strategy worked with T-Mobile losing market share in 2009–2010, leading up to the AT&T takeover bid.

But with the AT&T-fueled rejuvenation of T-Mobile, and Sprint's continuing challenges, it started to feel very much like Sprint was sliding into the valley of death, with T-Mobile climbing over us into a secure #3 position. A different strategy was required.

Pursuing a Sprint-T-Mobile Merger

IT BECAME APPARENT that Sprint's best hope was to combine with T-Mobile. Although T-Mo had tremendous momentum, the combination would also serve their interests well. Even in those first discussions in 2014 when Sprint would be the controlling party, we recognized that the T-Mobile brand was stronger than Sprint's, T-Mobile's GSM network had advantages over Sprint's CDMA network, and many of the key T-Mobile managers would be critical to the success of the venture. Masa Son's management style was very well

---

6. https://medium.com/clearpurpose/surviving-the-sprint-nextel-stragedy-part-2-b1338681b714

7. https://www.youtube.com/watch?v=nvTzJ4jlluA

aligned with T-Mobile's uncarrier strategy, and Masa positioned with regulators that the combined company would turbocharge that strategy, specifically fixing what is broken[8] with broadband services in the U.S.

Unfortunately, regulators didn't buy it. They were too proud of the positive impact their rejection of the AT&T/T-Mobile tie-up had made on the competitive environment and were dead-set against reducing the national market from four to three players. Their lack of support scrapped attempts to combine in 2014 and 2017, but thanks in part to a new regulatory administration and Sprint's unsustainable future, the deal struck between the companies in 2018 was finally approved by federal regulators late in 2019 and received final approvals earlier this year.

Is this Sprint Nextel All Over Again?

FOR THOSE WHO LIVED through the Sprint Nextel merger, the big fear is that this merger will result in the same outcome. Just like before, this merger is largely driven by the need to get bigger to eliminate the scale-driven advantages that Verizon and AT&T enjoy. Just like before, the new company has committed to maintaining a secondary headquarters in Overland Park, Kansas, and maintaining a large employee base there.

So, what's different?

For starters, this merged company has a strategy. Sprint's assets actually do enable "turbocharging" T-Mobile's uncarrier strategy. Sprint's 2.5 GHz spectrum is very complementary to T-Mobile's spectrum portfolio for offering 5G services. While T-Mobile has made strong progress in the past several years with small-to-medium business customers, Sprint's large business customer base and organization will enable the combined company to aggressively target a market dominated by the bigger Bells

---

8. https://www.cnet.com/news/softbanks-masa-son-on-why-sprint-needs-t-mobile/

with uncarrier offers. And then there's the residential broadband market that the company has promised to attack with uncarrier gusto. Bottom line, the addition of Sprint's capabilities and assets significantly strengthens T-Mobile's proven strategy in its core market and enables it to extend that strategy into new business and consumer markets.

But even more importantly, this is not positioned as a merger of equals. The executive team in Bellevue, Washington is calling the shots. Sure, they have tapped into the Sprint executive and management talent pool, but I don't expect the kind of internal culture wars that we experienced after the Sprint Nextel merger.

With a clear direction forward and an unconflicted command and control structure, I expect smooth sailing ahead for this worthy successor to Sprint. Godspeed and full speed ahead!

The Finish Line

─────

With this chapter we finally reach the end of the story, the end of the history of Sprint. That's not to say that we've seen the end of the impact that this great company will have on the telecom and wireless industries, or on society as a whole.

The world is a better place because Cleyson and Jacob Brown took entrepreneurial risks in Abilene, Kansas and decided that stringing up telephone lines on their electric poles might be a good idea. And because people like Skip Scupin and Carl Spaid invested their lives into growing United Telephone into a business big enough to make big bets. And because people like Paul Henson and Bill Esrey were willing to look beyond how the industry had always operated, to see how new technologies and new ways of using telecommunications could improve people's lives. And because thousands of smart and hard working people turned those visions into reality.

Arguably, United's decision to build a nationwide fiber optic network pushed the rest of the industry to move from analog to digital and accelerated the development and expansion of the Internet. Arguably, Sprint's decision to build a nationwide PCS wireless network spurred on the build-out of nationwide wireless data networks that we and our smartphones now take for granted. Arguably, Sprint/Clearwire's 4G network and Sprint's wholesale partnerships with trailblazers like Amazon for their Kindle eBook laid the foundation for today's Internet of Things. In an industry that has long suffered from complacency and conservatism, it's hard to argue against the impact that Sprint's innovations have had.

But Sprint's impact doesn't end now with the acquisition of a company or the disappearance of a brand. Tens of thousands of employees have passed through Sprint's doors and been changed for the better. We are smarter, more innovative, better connected, and more experienced than we were before, and we aren't sitting still. Seeing many people, once at Sprint, now scattered across the industry and beyond, gives me a strong sense for the impact Sprint has had, is having, and will continue to have into the future.

To give just a glimpse into the company's visionary leadership in creating the world in which we now live, let me one more time quote former United and Sprint chairman Paul Henson. These quotes come from a speech[1] he made on November 26, 1979 to the Midwest Research Institute:

"The first real test of our technical and marketing abilities will be to take computer and telephone technology and harness it for the benefit of the business and professional office. Over the last ten years, the productivity of the factory worker has gone up about 83%. In those same ten years, the productivity of the office worker has gone up only 4%. You all know, and probably utilize, the time-honored practice of dictation with a secretary typing the letter after it has been drafted a couple of times and you have done some editing. I haven't seen recent figures on the cost of the dictation-typing process, but it's got to be in the neighborhood of $10 a letter. The Postal Service says that by 1985 the first class stamp will cost between 25¢ and 35¢. Xerox claims that current facsimile machines can transmit within a matter of seconds an 8.5" x 11" page of paper for 4¢ a page. IBM asserts that its new satellite business system can do the same job in a matter of seconds for only 3¢."

"You may not realize that only 28% of all business calls are completed to the intended person on the first attempt. We in the telephone business

---

1. https://shsmo.org/sites/default/files/pdfs/kansas-city/mcp/Henson-11-26-79.pdf

don't talk about that a great deal since it represents a horrible waste of time and effort. We could solve that problem by sending hard copy, instantly, to the desk of somebody you want to communicate with – after you've tried your call, of course. If you don't find your party in or available to talk, hit another button and have the hard copy transmitted. We're working on it!"

"Telecommunications has made a great contribution to quality of life in America, as well as to the general economy. We now are on the threshold of advances which will have significant impact on the nation and on the midcontinent region. These advances will come about by a combination of the technologies which are developing in both the computer and telecommunications industries. It promises to be a very exciting business, one which will grow at rates at least twice those of the general economy. New technologies will impact both the electronic and print media. Technology and economics will change the way we do business. Properly applied, computer/communications can do much to improve productivity in the office and the factory, while providing services to the home which will do much to improve the quality of life."

"It is true that direct satellite-to-home transmissions are now technologically feasible and may soon be economically feasible. The major constraint is programming. How many old movies or obscure sporting events do you want to see? However, great strides are being made in the 'software' or programming aspects of cable TV. The QUBE system in Columbus, Ohio, is a case in point. Warner Communications and other such companies are experimenting with some very innovative two-way services using regular coaxial cable. While present costs are high, there's little doubt that two-way video systems are just around the corner. Direct home merchandising is a good example. The stores in Columbus, Ohio, are showing the housewife what styles they have in the ladies ready-to-wear department. Via a pushbutton telephone, the housewife can order a dress, have it delivered, and be billed for it. This is just the beginning."

"The costs of fiber optics and broadband transmission systems are coming down rapidly. It is anticipated that a single facility to the home and office will eventually provide all communications services – telephone, alarm, remote metering, energy control, educational or entertainment video, polling services, direct merchandising, library retrieval – you name it. Any kind of information you want in your home or office can ultimately be provided over a single facility at a low incremental cost."

The people of Sprint have envisioned the future and made it reality. Over the coming years, you will continue to see the people of Sprint asking smart questions and changing the world for the better. They may not be wearing a Sprint logo, but if you look closely at their LinkedIn profile and the impact they are making, you will know where they came from.

Don't miss out!

Visit the website below and you can sign up to receive emails whenever Russell McGuire publishes a new book. There's no charge and no obligation.

https://books2read.com/r/B-A-NWOL-BEPHB

BOOKS 2 READ

Connecting independent readers to independent writers.

About the Author

Russ McGuire is a trusted advisor with proven strategic insights. He has been blessed to serve as an executive in Fortune 500 companies, found technology startups, be awarded technology patents, author a book and contribute to others, write dozens of articles for various publications, and speak at many conferences. More importantly, he's a husband and father who cares about people, and he's a committed Christian who operates with integrity and believes in doing what is right.

Read more at sdgstrategy.com.

 **SDG STRATEGY**

About the Publisher

SDG Strategy helps faith-driven leaders of tech-driven startups with the hard decisions they face everyday.

Building and growing a startup is hard. Technology is always evolving, but your faith needs to be grounded in the unchanging truths and promises of scripture. As a leader in a dynamic environment, everyone is looking to you for all kinds of decisions, and few of them are easy. Russ McGuire has been there. Russ brings tools, methodologies, skills, and lessons learned from working with dozens of technology companies over 30+ years.

SDG offers live consultations, ongoing coaching, strategy lab workshops, online tools, and educational content. Visit us at http://sdgstrategy.com to learn more or to schedule a free 30 minute consultation.

www.ingramcontent.com/pod-product-compliance
Lightning Source LLC
Chambersburg PA
CBHW022125170526
45157CB00004B/1763